Housing
123

充足的食物

Plenty of Food

Gunter Pauli

[比] 冈特·鲍利　著

[哥伦] 凯瑟琳娜·巴赫　绘

唐继荣　张晓蕾　译

上海远东出版社

丛书编委会

主　任：田成川

副主任：闫世东　林　玉

委　员：李原原　祝真旭　曾红鹰　靳增江　史国鹏

　　　　梁雅丽　孟小红　郑循如　陈　卫　任泽林

　　　　薛　梅　朱智翔　柳志清　冯　缨　齐晓江

　　　　朱习文　毕春萍　彭　勇

特别感谢以下热心人士对童书工作的支持：

匡志强　宋小华　解　东　厉　云　李　婧　庞英元

李　阳　梁婧婧　刘　丹　冯家宝　熊彩虹　罗淑怡

旷　婉　王靖雯　廖清州　王怡然　王　征　邵　杰

陈强林　陈　果　罗　佳　闫　艳　谢　露　张修博

陈梦竹　刘　灿　李　丹　郭　雯　戴　虹

目录

充足的食物　　　　4

你知道吗？　　　　22

想一想　　　　　　26

自己动手！　　　　27

学科知识　　　　　28

情感智慧　　　　　29

艺术　　　　　　　29

思维拓展　　　　　30

动手能力　　　　　30

故事灵感来自　　　31

Contents

Plenty of Food　　　　　　4

Did You Know?　　　　　22

Think About It　　　　　26

Do It Yourself!　　　　　27

Academic Knowledge　　28

Emotional Intelligence　29

The Arts　　　　　　　29

Systems:
Making the Connections　30

Capacity to Implement　30

This Fable Is Inspired by　31

大熊猫夫妇坐在一片小竹林里，正享受着他们的美食。大熊猫爸爸却开心不起来，因为他能看到在竹林的尽头，人们正越过边界，建造房屋，开垦农田。

"我想知道我们还能在这片仅存的小小竹林里生存多久。"他咕哝着，没敢太大声地泄露出自己的担忧。

"为什么不开始改变你的饮食习惯呢?"一只待在附近的蟑螂开口问。

A pair of pandas is sitting in a small patch of bamboo forest, enjoying their meal. The panda father is worried however, as he can see where the forest ends, people building houses and farming beyond the boundary.

"I wonder how much longer we can survive off this little patch of left-over bamboo forest," he mumbles, not wanting to share his concerns too loudly.

"Why don't you start changing your diet?" asks a cockroach, sitting nearby.

大熊猫夫妇坐在一片小竹林里……

A pair of pandas in a bamboo forest ...

……地球上最受鄙视的动物。

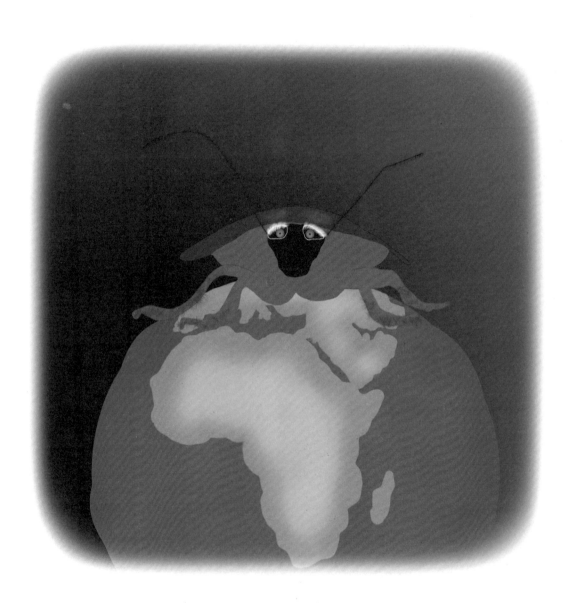

... the most despised animal on Earth.

"瞧瞧谁在说话呢！你可是地球上最受鄙视的动物啊！"

"也许吧！我猜，吸血的蜱虫可能比我更遭人讨厌。不过你要明白，讨厌我也不会让你得到更多的竹子。"

"你有办法让我和我的家人获得另一片可以生活的竹林吗？"

"Look who is talking! You are the most despised animal on Earth."

"Maybe! I guess that the blood-sucking tick may even be more disliked than me. But you know, disliking me does not get you more bamboo."

"Are you able to get me and my family another bamboo forest to live in?"

"抱歉，让你失望了，我没这本事！或者你可以试着吃点别的——比如这些盛开在竹林里的蝎尾蕉。"

"蝎尾蕉？多美丽的花朵呀！不过，它们不是给我吃的。"

"你以为花开在那里，只是为了让人欣赏它们的美丽吗？"

"Sorry to have to disappoint you, but I can't. Perhaps try eating something else – like these wonderful heliconias that grow in the forest."

"Heliconias? They are beautiful flowers but not for me to eat."

"Do you think flowers are there only to look beautiful?"

... wonderful heliconias ...

......鲜嫩的、含木质茎的竹子。

... young, woody bamboo.

"不，完全不是这个意思！只是蝎尾蕉不符合我的饮食习惯。我需要竹子，而且不是什么竹子都可以——我需要的是鲜嫩的、含木质茎的竹子。"

"好吧。那么你最好跟人类谈谈。"蟑螂建议道。

"哦，人们对他们的所作所为再清楚不过，因此现在我们将拥有一个公园，那儿适合我们生活，而且有充足的竹子。但是，那个公园面积很小，所以很多朋友及家人只得移居动物园，在那里他们能吃到那些我们最喜欢的竹子。"

"No, not at all, but it does not fit my diet. I need bamboo, and not just any bamboo. I need a specific type of young, woody bamboo."

"Well, you better talk to the people then," Cockroach suggests.

"Oh, they realise all too well what they have done. So now we will have a park where we can live and have plenty of bamboo. But it will be small, so many friends and family will have to emigrate to zoos, where they will be fed our preferred bamboo."

"既然一切都安排妥当了，你为什么还如此担忧呢？"

"我宁可所有的大熊猫都能在野生竹林中生生不息，但这似乎是一个不切实际的梦想。简单地说，人太多了，生存空间不足以容纳那么多人类和我们。"

"我不认为人太多呀。"蟑螂马上提出反对。

"So why are you so worried when everything is arranged?"

"I would have preferred that all pandas could thrive in wild bamboo forests but that seems an impossible dream. There are simply too many people and there is not enough space for them and us."

"I don't think there are too many people," Cockroach quickly responds.

······野生竹林······

... wild bamboo forests ...

……这么多肮脏的地方躲藏……

... so many dirty places to hide in ...

"啊？但人们肯定认为蟑螂太多了！"

"这只能怪他们自己了！只要他们还是随处留下这么多食物给我们吃，留下这么多肮脏的地方让我们躲藏，我们就无处不在。"

"我知道在人们的扑杀清单上，你们位列榜首，没有任何其他动物可以与你们比，然而也没有其他动物能像你们这样活得如此逍遥。"大熊猫说。

"Ah, but people certainly think that there are too many cockroaches."

"They only have themselves to blame! As long as they leave so much food around for us to eat, and leave us with so many dirty places to hide in, we will be around."

"I do not know any other animal that is so high up on their death wish list – and yet no other animal survives so well," Panda says.

"我们甚至在广岛原子弹爆炸中幸存了下来！你知道吗？我们蟑螂和你喜欢吃的竹子是在毁灭性的核爆炸之后最早出现的生命迹象！"

"你们是怎么做到的？"

"我们身上有一层起保护作用的蜡，可以防止我们受到伤害。另一件事也很重要：我们一直在调整饮食习惯，可以吃周围的任何东西。"

"We even survived the nuclear bomb in Hiroshima! Do you know that us cockroaches, and the bamboo you like to eat, were the firstsigns of life after that devastating nuclear explosion?"

"How did you do that?"

"We have a protective layer of wax that keeps us from being harmed. And another thing: we eat whatever is around, changing our diet all the time."

我们甚至在原子弹爆炸中幸存了下来……

We even survived the nuclear bomb ...

你并没有给我太多的希望……

you are not giving me much hope ...

"哦，我永远也不要那样做！"

"为了适应环境，从吃肉到吃竹子，你们大熊猫的饮食习惯在以前已经改变过一次。如果你们不准备再次做出改变，那么在这个世界上可能生存不了多久。"

"你并没有给我太多的希望。"大熊猫叹息道。

"瞧，要是你们还想在地球上再生存100万年，那就做出调整吧。"

"Oh, I would never be able to do that."

"You pandas have already changed your diet once before, from meat to bamboo because you needed to. If you are not ready to change again, you may not survive long in this world."

"You are not giving me much hope," Panda sighs.

"Look, if you want to be around for another million years, then adjust."

"是招人喜欢、受保护并被照顾，还是保持我们的生存方式而不在乎人们是否喜欢？或许我们必须在这两者之间做出选择。"

"确实如此！你们要下定决心：做你们过去一直在做的事情，或者为了生存并代代相传去做必须要做的事情。"

……这仅仅是开始！……

"Maybe we have to choose between being loved, protected and cared for, or keep our ways and not care if people like."
"Indeed! Make up your mind: do what you have always done, or do what is needed to survive and thrive throughout time."
... AND IT HAS ONLY JUST BEGUN! ...

……这仅仅是开始！……

... AND IT HAS ONLY JUST BEGUN! ..

大熊猫的食物几乎全部由竹叶、竹笋和竹竿组成。由于这些植物中所含营养很少，大熊猫每天不得不吃掉多达 40 千克的竹子。大熊猫只能消化所吃食物的 17%。

The panda's diet consists almost entirely of bamboo leaves, shoots and stems. As these plants have little nutrition, pandas have to eat up to 40 kg of bamboo per day. Pandas only digest 17% of what they eat.

大熊猫拥有熊科动物的消化系统，但食物组成中 99% 为竹类。所以它们每天要花费 14 个小时吃东西才能获得足够的营养。大熊猫的食物组成中只有 1% 是诸如小型啮齿类动物的肉类以及包括水果在内的其他植物。

Pandas have the digestion system of the bear family, yet 99% of their foods are bamboo. Therefore, they eat for up to 14 hours per day to get sufficient nutrition. Just one percent of their diet consists of meats such as small rodents, and other plants including fruits.

2-3 mth

It is only when a female panda has obtained sufficient nutrients, including calcium needed for lactation, that a panda embryo starts growing. The panda has only a 2-3 month gestation period, compared to a minimum of 6 months for other bears.

只有当雌性大熊猫获得充分的营养——包括哺乳所需的钙质，大熊猫胚胎才会开始生长。大熊猫的妊娠期只有2—3个月，而其他熊类的妊娠期至少有6个月。

100g

A baby panda only weighs about 100 grams at birth. Its tiny size is due to the limited amount of available nutrition from mother panda.

大熊猫宝宝出生时体重大约只有100克。它的体型小是由于母体的营养供给有限。

大熊猫的手掌看上去有6根手指。除了像其他食肉动物那样正常的五根手指，大熊猫前肢的手掌还有一根额外的"拇指"。这根"拇指"实则是桡侧腕骨在进化中变形所致，有助于大熊猫在进食时抓握竹子。

It looks like that panda's palm of hand has six fingers. In addition to normal five fingers just as other carnivores, panda's paw on forelimb has an additional "thumb". The "thumb" actually formed from a modified radial carpal bone during evolution, which helps to hold bamboo while eating.

In addition to normal colour of black and white, there are also brown and white pandas in Qinling Mountain, Shaanxi Province.

除了正常黑白体色的大熊猫，在陕西秦岭还有棕白色的大熊猫。

By the end of 2013, there were 1864 wild pandas in the forests of Sichuan, Shaanxi and Gansu in China. At the same time, there are over 600 pandas in captivity in the world.

截至 2013 年年底，中国有 1864 只野生大熊猫，分布于中国四川、陕西和甘肃的森林中。同时，全世界目前还人工饲养了超过 600 只大熊猫。

Pandas evolved specific survival skills, which allows them to exist more than eight million years. Compared to other animals that lived at same period but extincted at last, the panda is survivor of natural selection. Therefore, the panda is also called living fossil.

大熊猫进化出独特的生存技能，在地球上生存了超过 800 万年。与其同期的动物大多已灭绝，大熊猫是自然选择的幸存者，有"活化石"之称。

Think About It

想一想

Would you be ready to switch to an all vegetarian diet?

你准备好把饮食转变成全素食了吗?

If you cannot get your favourite food, would you rather starve to death, or change your diet?

如果你不能得到你喜欢的食物,你是宁愿饿死,还是改变你的饮食习惯?

Do you want to be popular, and do and be what people want you to be?

你想成为受欢迎的人吗?你愿意为此做人们想要你做的事、成为人们想要你成为的人吗?

Is the cockroach on your death wish list?

蟑螂在你的扑杀清单上吗?

Do It Yourself!
自己动手！

Are more people following a vegetarian diet? What if you were to have one meat-free day a week, or a vegetarian week once a month? How would you substitute animal products with plant-based products? Let's start preparing a delicious dessert: chocolate mousse using avocado. Ask your mum or grandma to help you combine two ripe avocados with 40g of carob powder in a blender. You can add a little milk to make it more runny. It is safe for you to prepare this as it does not require any cooking or baking.

有更多的人在奉行素食主义吗？如果你每星期都有一天不吃肉，或者每个月都有一周食素，结果会怎么样呢？你会如何用植物产品替代动物产品？让我们动手准备一道美味的甜点：用牛油果做的巧克力慕斯。请你的妈妈或奶奶帮忙，在搅拌器中把两个成熟的牛油果和 40 克角豆荚粉混合。你可以加一点牛奶，让它更湿润。慕斯不需要烹饪或烘烤，你可以安全地制作这道甜点。

学科知识

Academic Knowledge

生物学	蝎尾蕉属植物喜欢在竹林里生长；地衣可以从岩石中吸收矿物质，这使其具有生物可利用性；竹类的生物多样性和大熊猫喜爱的竹子种类；中医认为，富含维生素的玫瑰是花中之王，可以降低患心脏病的风险；茉莉花有抗癌和抗病毒的功效。
化 学	哺乳需要钙质；岩石中矿物质的含量；可食用的花都富含酚类和抗氧化剂；三色堇颜色鲜艳，富含钾元素。
物 理	核辐射；蜡可以作为防辐射的保护层；竹子不仅为大熊猫提供食物，还为它们创造了更清凉的环境。
工程学	工程师可以采取多种方式来保护生物的栖息地，防止物种灭绝；材料工程师可以开发新材料来模拟木材的特性，以此取代热带雨林的木材；化学工程师可以研发各种手段减少海星对珊瑚礁的不利影响；环境工程师可以研发抑制全球变暖的方法，虽然这会对珊瑚礁带来不利影响；农业工程师可以发展农作技术，抑制土壤养分枯竭并保持土壤健康。
经济学	经济的韧性依赖生物多样性；生物多样性是大自然灵感和创新的源泉，它守卫着大自然中许多巧妙的发展，而这些发展是数百万年来进化、共生和适应的结果；对生物多样性进行明码标价；印度因三种秃鹫的几近灭绝而付出近240亿美元的代价。
伦理学	既然每个物种都承担特定的功能，人们怎么能仅仅因为他们不喜欢就把物种列入扑杀清单呢？
历 史	数百万年前，大熊猫成为素食者；1966年，世界自然基金会（WWF）把大熊猫作为它的会徽，这极大地提高了它的公众关注度，有助于筹款和保护目标的实现。
地 理	大熊猫在中国西南和西北地区的栖息地；广岛。
数 学	反弹效应：尽管越来越多的人减少肉食，但由于中产阶级人数的增加，整体消费肉食的水平仍在提高；竹子生长激发了一个超级公式的诞生，即吉利斯公式。
生活方式	减少肉食的趋势；为了使人们在动物的栖息地之外也能看到全世界的动物而建造动物园；用来装饰婚礼的牡丹花可以食用，有缓解抑郁的效果。
社会学	迁移到其他栖息地的需要以及适应新的生存环境的能力。
心理学	人们会对一些自己怎么也不能改变的事情感到内疚；能够灵活变通，适应我们周围变化的能力；我们为什么会不喜欢某些人和事；因为有些事情我们不了解，即使它可能是件好事，我们仍然会拒绝接受；把责任推到别人身上；焦虑的心理状态。
系统论	肉类消费减少会对土地利用和气候变化造成影响；环境容纳量的概念：在有限的土地上能维持种群生存的大熊猫数量；需要创造一个有利于生活的环境。

情感智慧
Emotional Intelligence

大熊猫

大熊猫很忧虑，有倾诉的欲望，但又不想造成别人的困扰。他默许了蟑螂的陪伴，但语气中带着一种优越感，可一旦意识到蟑螂可能有一个新的解决方案，语气就变了。大熊猫思想保守，并不准备去寻找其他办法。他把搬进一个公园定居下来作为自身的解决方案，但由于许多同伴和亲人要移居到其他地方，他又备感压力。他用居高临下的口气与蟑螂说话，发泄着他的沮丧。而核爆炸之后竹子和蟑螂的复苏给了他启发，他想要了解更多。大熊猫固执但缺乏自信。更糟糕的是，他失去了希望。他知道得由他自己做出选择。

蟑　螂

蟑螂质疑大熊猫的态度。面对大熊猫的贬低，她仍毫无顾忌地高谈阔论。蟑螂主张大熊猫应该改变自己的饮食习惯，并坚持认为改变是必需的。当大熊猫描述一个并不完善的解决方案时，蟑螂督促他从自身寻找答案。面对大熊猫的居高临下，蟑螂却并不在意，并表明了她在极端逆境时的力量。她提醒大熊猫，他已经经历了一次根本性的转变，还可以再经历一次这样的转变，这为大熊猫下定决心去做生存所需的事情提供了动机。

艺术
The Arts

竹林是许多艺术家绘画和作诗的灵感来源。找一些以竹子为主题的艺术作品，同时找出一两首赞美竹林的美丽、清新、静谧和安详氛围的诗。阅读和欣赏过一些关于竹林、大熊猫的诗、画作和图片后，自己也来写一首诗吧。你的这首诗可以是关于你想成为什么样的人、其间遇到的困难以及你如何调整自己去达成需要，还可以包含他人的关注与期许。

思维拓展
Systems: Making the Connections

地球正在经历不断的变化。没有什么是一成不变的，而生存往往取决于适应能力。自然正处在进化的道路上，像陨石和火山爆发这样的重大影响已经迫使现存物种为了生存和繁盛而发生转变，人类这个物种的出现只是地球生命的诸多根本影响力之一。大熊猫的栖息地从低海拔向高海拔迁徙，该地区严重缺乏动物蛋白，但富含植物蛋白，尤其是竹子。大熊猫选择不与其他食肉兽类竞争，而是改变自身的食性。尽管大熊猫的消化道当时根本不能适应这种变化，甚至至今依然不太适应，但为了生存也只得做出改变，为此大熊猫付出了巨大的代价。大自然的变化，对大熊猫的生存提出了严峻的考验，要么被大自然淘汰，要么选择做出改变来适应环境。后者也许不是大熊猫想要做的，但它们只有这样做，才能随着时间推移，继续生存和繁衍。

动手能力
Capacity to Implement

世界上有许多濒危物种，它们似乎没有足够的栖息地，这也意味着没有充足的食物。随着人类的种植活动占用了更多土地来生产单一品种的食物，同时随着生产性土地被水泥城市和道路覆盖，人类正在排挤其他生物。因此，与其预期其他物种灭绝，或者调整以适应这种以人类为中心的新的生活设计模式，同时只在动物园里饲养少数个体供展览和研究，还不如问自己这个问题："为了跟上自然演化的步伐，在我们的有生之年，我们可以做出什么样的改变？"这些改变并不一定要立即发生，但一定要展示出那些你在有生之年相信能够而且必须做到的改变。要想动员许多其他人加入你的行动，成功的关键是立即在正确的方向采取一些步骤。因此，需要确定一些容易达成、能非常快速地显现冲击力的行为。毕竟，在具备要前往何处和如何到达的清晰愿景的同时，生存取决于这些策略的速度和尺度以及采取没有任何进一步延误的行为。

故事灵感来自
This Fable Is Inspired by

魏辅文
Wei Fuwen

魏辅文是一位保护生物学家。他在西华师范大学先后获得生物学学士学位和野生动物生态学硕士学位,随后在中国科学院动物研究所获得保护生物学博士学位。他现在是该动物研究所的副所长和研究员,并于2001年获得中国国家杰出青年科学基金。他的工作集中在濒危物种的过去、现在和未来的生存状况,特别关注中国的大熊猫和小熊猫。他研究和分析了大熊猫的进化和种群数量统计,并领衔制定大熊猫的生存策略。他是世界自然保护联盟物种生存委员会专家组成员,并担任中华人民共和国濒危物种科学委员会副主任。

图书在版编目（CIP）数据

冈特生态童书.第四辑：修订版：全36册：汉英对照 /
（比）冈特·鲍利著；（哥伦）凯瑟琳娜·巴赫绘；
何家振等译.—上海：上海远东出版社，2023
书名原文：Gunter's Fables
ISBN 978-7-5476-1931-5

Ⅰ.①冈… Ⅱ.①冈…②凯…③何… Ⅲ.①生态环
境–环境保护–儿童读物—汉、英 Ⅳ.①X171.1–49

中国国家版本馆CIP数据核字（2023）第120983号
著作权合同登记号图字09-2023-0612号

策　　划 张　蓉
责任编辑 张君钦
封面设计 魏　来　李　廉

冈特生态童书
充足的食物
[比]冈特·鲍利　著
[哥伦]凯瑟琳娜·巴赫　绘
唐继荣　张晓蕾　译

记得要和身边的小朋友分享环保知识哦！
八喜冰淇淋祝你成为环保小使者！